Henry W. Acland

Synopsis of the Pathological Series in the Oxford Museum

Provisionally arranged for the use of students after the plan of the

Hunterian Collection, and chiefly under the divisions of the Hunterian

catalogue

Henry W. Acland

Synopsis of the Pathological Series in the Oxford Museum
Provisionally arranged for the use of students after the plan of the Hunterian
Collection, and chiefly under the divisions of the Hunterian catalogue

ISBN/EAN: 9783337424138

Printed in Europe, USA, Canada, Australia, Japan

Cover: Foto ©berggeist007 / pixelio.de

More available books at **www.hansebooks.com**

SYNOPSIS

OF

THE PATHOLOGICAL SERIES

IN THE

OXFORD MUSEUM.

𝔏𝔬𝔫𝔡𝔬𝔫

MACMILLAN AND CO.

PUBLISHERS TO THE UNIVERSITY OF

𝔒𝔵𝔣𝔬𝔯𝔡

SYNOPS████

OF

THE PATHOLOGIC█████████

IN THE

OXFORD MUS█████

PROVISIONALLY ARRANGED FOR THE USE OF ST████
OF THE HUNTERIAN COLLE████
AND CHIEFLY UNDER THE DIVISIONS OF THE H████

𝔒𝔵𝔣𝔬𝔯𝔡:

AT THE CLARENDON PRESS.

M.DCCC.LXVII.

CONTENTS.

NOTICE TO STUDENTS.

The student of this Synopsis is recommended, first, to read with close attention the Introduction ; and secondly, the summary of the first subject—Hypertrophy. He should next examine the preparations referred to under the head Hypertrophy, with the aid of Vol. I of the manuscript Catalogue, which is to be found at the commencement of the series of General Pathology.

He will then study the references which follow. All the works so referred to are in the Pathological Department, or in the Library. He will pursue a similar course with the succeeding sections.

After this survey of the General Pathology, he may examine the Special Pathology in the same manner.

A detailed manuscript Catalogue of the General and of the Special Pathology, in sixteen volumes, is placed in the compartments to which the volumes respectively belong.

The first numbers on the white labels upon the preparations refer to the MS. Catalogue in English. The second numbers refer to Schroeder Van der Kolk's Catalogue in Latin. Green labels are affixed to those preparations which specially illustrate points in General Pathology, but they are described in their place under the first number.

In the department is a small collection of instruments of diagnosis. The beginner is recommended to learn the use of these, and if he have the opportunity, to use them practically. If he desire either of these he is to apply to the Regius Professor of Medicine.

INTRODUCTION.

THE student of biology has completed but a portion of his task when, with the help of an extended physiological series, he has learnt the general principles, and some of the details, of the genesis of living beings: of their growth, and the means of their maintenance; and of their relation to each other in time and in space.

There remain for him biological questions not less intricate nor less momentous. Of these, the first is, 'What are the laws of decay?'

Perpetual change of matter, or the capacity for change, is the law of life. The accretion of new matter, the extrusion of old matter, are among the means which are going on almost without cessation in the minuter parts of every living being. The parts of an individual, therefore, are formed, completed, and discharged, in constant succession, while the individual, to all appearance, remains unaltered.

This unceasing passage of matter through the individual is therefore an essential, and, as far as we know, a necessary condition of its active life—of its health.

But an epoch comes, very different in different species, and variable in the same species, when these normal changes no longer take place, and when (just as the parts of the whole being had come to an end through millions of successive generations of individual parts) the individual as a whole comes to a natural end. So entirely is this the law of life, that the term 'natural decay' expresses a fact appreciated and accepted by the general sense of mankind.

But beyond and beside the processes or modes by which the individual 'naturally' comes to an end, there are modes of decay which lead to an 'unnatural decay'—a premature death.

These modes of unnatural decay, or of premature death —that is, of destruction other than by the normal course —take place equally in the component parts of the individual and in the whole individual.

As the component parts of an individual are not all of equal importance to the total life, so many component parts may be destroyed in an abnormal manner, and the individual remain on the whole as long-lived, as far as we know, as though no such destruction of parts had occurred. A man may be long-lived though early bald. Yet the loss of any texture such as the hair, implies a deficiency (however slight) in some of the power of re-generation of that texture, however otherwise compensated. The want of re-generative power in some of the minutest parts of the organism may, however, unlike the case of the hair, be fatal to the individual. A premature degeneration of the component parts of the muscular texture, a premature or abnormal want of nutrition of the minute parts of the nervous centres, may, and in certain instances necessarily do, lead to more or less rapid destruction of the whole individual.

Just so some diseases imported into a race, by attacking the individuals of that race, might expunge, and in some cases have expunged, that race from the world, never to re-appear in it.

As then the biological student enquires into the facts and laws of formation of living beings in ordinary *physiological* investigations, so in the *pathological* department he has to enquire into the material operation ' præparantes ad mortem,' both usual and unusual.

Not that an absolute line of demarcation can be drawn between healthy and morbid processes. It has been just said that a living individual lives by or with constantly recurring death of his component parts. But this death of the parts proceeds in a certain way among the healthy, and in another way among the so-called unhealthy. To investigate this latter way is one special function of *pathology.*

But to understand the abnormal way in the whole, or in the details, it is manifestly necessary first to understand the normal tendency to an end.

Changes prematurely fatal to the organism may occur, which are inappreciable except through the microscope of a highly-skilled observer; while on the other hand extensive changes of even the most important organs may take place with but little disturbance to the life or comfort of their possessor.

Such changes, great and small, may be detected or may be undetected during life; they may be revealed after death, or, as is often the case, may even then be undetected, and perhaps undetectable.

Pathology, therefore, is not to be learnt by the biological student in the museum alone; it has to be also learnt in the more complex conditions of the individual who lives in an abnormal state.

But the *pathological collection* does show the termination of many abnormal processes—that is to say, extreme conditions which are inconsistent with life—and it shows much more: it shows eminently the early stages of collateral disorders which had not reached the culminating point of disturbance in their own class, and which were not the cause of death.

In this manner the whole history of morbid processes is being gradually reached. Life is brought to a close by the advanced disorder of one texture or organ, while simultaneously processes of early, perhaps innocuous, but still abnormal, change may be in progress in other parts.

But the pathological student has opened to him a series of changes wholly different from what have here been noted.

Life is not wholly resistance to decay, nor is it wholly a tendency to death. It is like the ceaseless action of the tidal sea, which, flowing and ebbing day by day, never the same, yet always like the same, silently and yet certainly hews the cliff to its predetermined fall.

Repair goes on, only not quite so steadily, as decay. Decay of the whole must in the end prevail, yet in the individual parts again and again repair exceeds decay. Again and again, moreover, fatal change is averted by conservative tendencies in the very pathological process.

The law of REPAIR, therefore, is a part of pathological study, just as much as is the law of death. Whether the fractured limb is restored by the aid of art, or whether the nautilus, rent on the rock, is repaired by the work of the mantle, the process of repair in each is subject-matter of pathological investigation.

Thus *pathology* is bound to *physiology* on the one side, and on the other to the *healing art.* It examines the departures from health, notes the limits of variation which are inconsistent with health, records in the disruptions and disturbances of morbid anatomy the variations which have produced permanent alterations of structure, whether they have or have not been fatal to life.

Pathology is GENERAL, or SPECIAL.

GENERAL PATHOLOGY treats of morbid processes which are common more or less to the whole frame—with excess, with diminution, with repair and reproduction, with the processes of inflammation and of mortification, and with out-growths and tumours, all which processes may occur in some form in all the textures which the anatomist and physiologist have described.

SPECIAL PATHOLOGY treats of the disorders to which each individual texture—as blood, muscle, cartilage, bone, and the like—is liable, with the derangements of the several organs of the body as composite machines.

Pathology is therefore the basis of the classification of diseases. It does not profess to show the causes of diseases, nor the way of preventing diseases, nor the way of relieving them. Its proper limitation in those directions is to show what are the effects of agents which can act injuriously on living bodies.

This limitation is one with a vast circumference; for, as the science advances, it will state what circumstances of climate and of race modify the effects of noxious agents on different persons.

So that pathology will gradually incorporate, not only the history of the changes which take place in any animals, or in man, in any one country or under any one condition, but gradually also the external circumstances and the internal constitution which regulate those changes.

Pathology, therefore, necessarily touches on the history of epidemic influences and epidemic cycles, seeking to show, for instance, whether a poison does or does not produce a more fatal or a less fatal effect at one period than at another, in one race rather than in another, as it does in one species and not in another; and this especially by noting the lesions respectively produced under the several modifying circumstances.

That the interests involved in pathological enquiry have a vastness more than co-extensive with the anatomical limitation of the human family, is apparent.

In the last half-century there has been shed a flood of light upon morbid changes which had been denied to previous ages. The means of physical, mechanical, and chemical enquiry during life, the more precise research after death, especially by the microscope, have altered the whole aspect of human knowledge in this direction. The changes produced by various agents on living structures, for good and for evil, remain to be more precisely noted.

A natural, as distinguished from an empirical, classification of morbid processes leading to death, is becoming more possible*, and may ultimately be attained. The correlations of disease, and the laws which regulate those correlations in a mechanism so intricate and so ever-varying as the human body, can hardly be otherwise than too complex for absolute classification into formulæ of universal application†.

It remains to state what is the special, though limited purpose, of the following few pages.

It appeared essential twenty years ago to extend the means of biological study in this University, both for the purpose of general national education, and for the fundamental scientific training of students for the medical profession. One of the first systematic steps was to lay the foundation of an extensive physiological series, which should contain the germs of every part of biological study.

Of a part of this attempt a brief account was published in 1853‡. That account generally set forth the elementary

---

* The classification of disease by the Royal College of Physicians of London will aid in no small degree this end, and the principle of decennial revision will give permanence to this great work. When the classification is generally accepted, it will be for consideration how far this collection can be adapted to it.

† The pathological student should peruse the papers of Dr. T. K. Chambers, entitled 'Decennium Pathologicum,' in the *British and Foreign Medico-Chirurgical Review*, vol. xi. April, 1853, and vol. xii. October, 1853.

‡ *Synopsis of the Physiological Series in the Christ Church Museum, arranged for the use of Students after the plan of the Hunterian Collection, and chiefly under the Divisions of the Hunterian Catalogue.* Oxford, 1853.

principles of physiological study, as a key to the
study of the healthy organism. It was not desirable
at that time to take notice of the pathological portion
of the collection, which, according to the opportunities
that offered, was slowly forming.

But it was pointed out* that the student 'must en-
deavour to understand what changes take place in the
living beings themselves; by what means they are formed;
what changes they undergo during this process of
formation; by what means they are nourished when
fully formed, in their several parts and in their whole;
through what combinations and by what means the matter
which has performed its work in the living being is
eliminated from it; and also to what extent these changes
are necessary in individual parts of the organism for the
maintenance of the whole; what forces—external to the
organism, or internal, whether physical or vital—are
engaged in maintaining this continuance of life; *how far
injuries and losses of parts are capable of repair,
and by what circumstances, and through what steps,
the cessation of these living phænomena, or death, ensues.'*

Since that time the Christ Church collection, the subject of the work just quoted, has been, with much liberality
and true public spirit, lent by the Dean and Chapter to
the University, and is transferred to the hands of the
first possessor of the newly-created physiological chair
—Professor Rolleston. Under his energetic and generous
administration, every portion of the nucleus committed
to his care is being increased, to more goodly propor-

* Physiological Synopsis, p. viii.

tions. Especially the anthropological part is undergoing more ample development. Though the broad basis of comparative anatomy is not neglected, Ethnology proper, in its widest sense of time and place, bids fair to occupy the attention which it should occupy in Oxford, a place devoted to the study of man's nature, of his past history, and his possible destiny.

As has been said above, the biological student, that is, the philosophical student of physiology, should, in his summary survey of man, seek to understand the ways 'præparantes ad mortem.' For this purpose he will pursue the course recommended in the Notice at page 6.

What is desirable for the philosophical student of biology as a branch of general culture, is a necessity for the student of the healing art, as one of the cornerstones of his profession. He, however, must do this work in combination with the study of symptoms during life, for the reasons alleged above*.

These then are the two classes for whose instruction this series is prepared. For the purpose of general culture the study is desirable as a complement to the history of growth and formation: for the purpose of the future physician or surgeon, it is necessary as an aid indispensable for the comprehension of curative processes, and of modes of death.

Before concluding this brief Introduction, something should be recorded concerning the small collection which it seeks to illustrate, and concerning the Synopsis itself.

* Pages 10, 11.

First, the collection consists of a few preparations procured by Professor Kidd from his professional brethren in this district. To increase these was one of the objects which I set before myself when the extended physiological series was commenced in 1845. The pressure of many duties, and the inherent difficulty of obtaining morbid specimens away from great centres of population, have made progress in this direction slow; but I desire to record the aid kindly given by many friends: by William Ormerod, Mr. Hitchings, Mr. Cleoburey, Mr. Parker (all once colleagues in the Radcliffe Infirmary, but, alas! long since deceased), by Dr. Jackson, Dr. Child, Mr. Hester, and Mr. F. Symonds, and to a great degree by Dr. Gray, Mr. Hussey and Mr. Briscoe, and now, also, by Mr. Winkfield.

To no one am I so indebted for the dissection of the older preparations, and the preservation of all, as to Mr. Charles Robertson, for fifteen years my valued assistant. His help is, even now, never wanting, when it can be given consistently with his other duties.

When the Lee's Readership in Anatomy was resigned by me in 1857, the Dean and Chapter added to their many favours that of giving to me personally the pathological collection, considering (as they were good enough to say) that as it had been created mainly by my labours and at my expense, to me it properly belonged. There was in my new department no apparatus of any description. The University has provided the present accommodation for it; and I have made it over to them.

Secondly, in 1865 information was given to me that the valuable private collection of the late Professor Schroeder van der Kolk, one certainly of the most noteworthy of European pathologists, was to be sold. I went to Holland to examine it, and obtained the whole of the pathological portion for Oxford, having the pleasure at the same time of purchasing the physiological part for Professor Humphry of Cambridge.

The two collections from Christ Church and from Utrecht are now incorporated, and form one series, arranged under the heads of the Hunterian Museum. There were manuscript catalogues of each.

These catalogues and the united collections were entrusted by me to Dr. Henry Tuckwell, now Physician to the Radcliffe Infirmary, and formerly a Radcliffe Travelling Fellow.

He has arranged and re-described these collections; made an abridged translation of Schroeder van der Kolk's MSS.; prepared MS. hand-catalogues of the several parts, and made several dissections of much interest. He has, further, prepared the Synopsis, in continuation of the plan of the Physiological Synopsis above alluded to, and now carries on the work by undertaking, at my request, to demonstrate the collection to any professional or non-professional members of the University, whose general and scientific education is sufficiently advanced to profit thereby.

These facts record better than many words how great are my obligations to my colleague, Dr. Tuckwell.

At some future period perchance, a Synopsis of the Special Pathology, as well as of other series bearing on General Medicine, may be prepared. Meanwhile the block is hewn out, and there remains but to fill up details. Should these pages meet the eye of any who can serve the progress of scientific education in Oxford, they may give real help by presenting to us good dissections, and especially microscopic preparations, in morbid anatomy. Some illustrations of the minute morbid changes of the nervous system are much required. A few microscopic preparations have been presented by Mr. Payne, Fellow of Magdalen College, Radcliffe Travelling Fellow.

Whatever will illustrate the pathology of vegetable or animal life, taken in the sense of this Introduction, whether drawings, plans, chemicals, or dissections, will at once find its natural place in the classification of the series, to the elucidation of which these pages are an instalment.

HENRY W. ACLAND.

OXFORD MUSEUM,
October 18, 1367.

I.

GENERAL PATHOLOGY.

GENERAL PATHOLOGY.

I. HYPERTROPHY.

TO understand the nature of excessive Nutrition, or Hypertrophy, is to understand the first link which binds Physiology and Pathology. Hypertrophy, in its *first* or simplest form, may be looked upon as a conservative physiological process for the purpose of remedying or counteracting the ill effects of disease. It seems to be a result of what Paget has called a " reserve-power" in nutrition, present in every healthy body. By virtue of it, and in cases of emergency, excessive nutrition and consequently excessive growth take place in a part that is excessively worked. The muscular tissue, organic as well as inorganic, affords the best illustrations of this. For instance: the left ventricle of the heart increases in power and size when the circulation is interfered with by obstruction in the aorta, by disease of the mitral valves, or by disease of the kidneys. The muscular coat of the urinary bladder may be in like manner thickened when the escape of urine is impeded by stricture of the urethra, or enlargement of the prostate gland. (*See Preparations* 1—4.)

From cases of genuine hypertrophy such as these, we pass to the consideration of a *second* class of hypertrophies, which, because, as far as we can now see, they serve no healthy purpose in the economy, but on the contrary are prejudicial, must be regarded as pathological rather than physiological changes. Such are enlargements of the thyroid and prostate gland, of the tonsils, spleen, etc. (*See Preparations in Special Pathological Series*, Nos. 544, 731, 750, 752.)

And *thirdly*, there is a class of still more strictly morbid

c

hypertrophies, which occupies a sort of middle ground between hypertrophies and "tumours," and which will be better considered when the latter are described.

Hypertrophy may be divided into *simple hypertrophy*, or hypertrophy by growth; and *numerical hypertrophy*, or hypertrophy by development. In the former the number of original elements in the part remains unchanged, their size alone being increased by an addition of homologous particles. In the latter (for which Virchow has proposed the name of "hyperplasia") not only are the elements enlarged but their number is also increased. It may be presumed that the two forms occur simultaneously in most instances of hypertrophy. Thus, in the enlargement which the uterus undergoes when fibrous tumours are developed in its walls, simple hypertrophy is shewn by an unusual development and growth of the pre-existing muscular fibres; numerical hypertrophy by an entirely new formation of fibres which did not before exist. The latter is, however, more fully exemplified in parts composed of cells (as the epidermis): here numerical increase predominates over simple growth.

Hypertrophy must be regarded as essentially an *active process* set in motion, for the most part, by the operation of an *irritant*. This irritation may be produced in a variety of ways:—

1. Simply "by the unusual exercise of a part in its healthy functions."
2. The irritant may be mechanical: instances of which are seen in the localised hypertrophies of epidermis, or callosities, occurring on the feet from pressure of boots, on the hands from friction caused by various crafts.
3. "By the accumulation in the blood of the particular materials which a part appropriates to its nutrition or in secretion:" as, for instance, of the constituents of the urine causing hypertrophy of one kidney when the other is atrophied or otherwise diseased.
4. By the action of a blood poison, e. g. hypertrophy of the spleen in fevers.

A part that is undergoing hypertrophy naturally requires an increased supply of blood or nutritive material. This unusual afflux is, as a rule, a secondary phenomenon; the blood is attracted by the irritated tissues. But there are cases in which it appears that excessive vascularity of a part is the primary cause of the hypertrophy;

as in the instance adduced by Hunter of the extraordinary growth
of the cock's spur when transplanted to the vascular comb. (*See Preparation* 21.)

*Hunterian Pathological Catalogue, Vol. I.; Paget's Lectures on
Surgical Pathology, 2nd Edition,* 1863, *p.* 47 *et sqq.; Rokitansky,
Pathological Anatomy, translated by the Sydenham Society,
Vol. I. p.* 37; *Lebert, Anat. Path. Vol. I. p.* 85; *Virchow, Handbuch
der Speciellen Path. und Therapie, Vol. I. p.* 326.

II. ATROPHY.

Atrophy, like hypertrophy, is, strictly speaking, a physiological
process, being the change to which all bodies sooner or later naturally
tend; and thus forming as truly a stage in the cycle of life as does
development or growth.   But, regarded from another point of view,
atrophy is the very opposite of hypertrophy.   For, whereas the latter
has been seen to indicate increased nutritive power, the former
marks a defect in that power, a retrogression from the even balance
of nutrition which we term health.

Atrophy may be divided into two forms, *simple atrophy*, in which
parts merely dwindle or decrease in size and power without undergoing
any absolute change in their constituent particles,—in other
words, waste : and *atrophy by degeneration*, in which parts do not
necessarily decrease in size, (nay, may even become larger,) but are
structurally deteriorated by the substitution of a lower or less highly
organised material for their own perfect tissue,—in other words, are
qualitatively rather than quantitatively changed.   Both these forms
are met with in the body's natural decay; nor is it right to regard
the latter, because it is called a degeneration, as less of a physiological
process than the former.

Simple atrophy is best studied in the emaciation or loss of fatty tissue in the old.

Degeneration manifests itself in a variety of forms, according to the nature of the material substituted for that which is proper to a part. Thus the material may be fat, and there ensues what is termed "fatty degeneration," the most common of all, and met with in all organs of the body: or it may be lime, giving rise to the "calcareous degeneration" most marked in the arteries: or pigment, causing the "pigmental degeneration," so frequently met with in the lungs of old people.

Although neither form of atrophy considered thus far is pathological, yet the second form, degeneration, is so intimately connected with disease, that it must, in certain conditions, be looked at as really a morbid process: and it would be an error to omit the study of degeneration as a pathological process, when occurring unnaturally in the earlier periods of life, because it is a usual cause of death at advanced years. Tumours and products of inflammation have a frequent tendency to undergo the same changes that have been already mentioned as varieties of degeneration, especially the fatty change.

Certain organs, as the liver, are liable to what is called "acute atrophy," in which the tissues are rapidly destroyed as such, and abundant fatty granules or molecules take their place.

Some of the causes of atrophy are the opposite of those which produce hypertrophy. 1st. Disuse or diminished exercise of a part in its healthy functions. 2nd. A diminution in the blood of the particular materials which a part appropriates in its nutrition. 3rd. A diminished supply, *a.* of blood, *b.* of nerve force, (both of which may be consequences as well as causes of atrophy). 4th. Constant pressure; this causes atrophy just as occasional or remitting pressure produces hypertrophy.

<table>
<tr><td>Wall-Cases under Gallery No. 7.</td><td>Nos. in Catalogue.</td></tr>
<tr><td>Atrophy of nerves from disuse ....</td><td>5—11</td></tr>
<tr><td>Atrophy a result of constant pressure ....</td><td>12, 13</td></tr>
<tr><td>Atrophy of kidney (probably congenital) ....</td><td>14</td></tr>
</table>

See *Hunterian Cat. Vol. I.; Paget's Lect.* 1863, *p.* 69; *Rokitansky, Vol. I.* 49; *Virchow's Handbuch, Vol. I. p.* 303.

III. REPRODUCTION AND REPAIR.

The power of recovering from the effects of injury or disease, whether manifested in its most perfect form of reproduction or in its less perfect form of repair, appears to be possessed by all bodies that have a definite form.

Instances of reconstruction *of the whole body* from a fragment of the old one must be looked for only in the lowest animals, such as the polypes*. Reproduction *of entire limbs* is common enough in animals as high up in the scale as the salamanders. Trustworthy cases are also recorded, in the human embryo, of reproduction of perfect fingers on the stumps of arms that have been amputated in utero. In man and the other mammalia, certain tissues of low organization and low chemical character (as epidermis, connective-tissue and bone) are readily reproduced: but the higher tissues are with difficulty restored. New nerve fibres, and perhaps cells, are formed in certain cases where a nerve has been cleanly divided without much injury to, or bruising of, the parts; but new muscular fibres are, as far as is yet known, never formed. The rule is, that in these higher tissues a breach of continuity is mended by the interposition of a lower tissue [the connective-tissue], so that the part seldom recovers its pristine perfection.

The study of Repair is best followed out in injuries, viz. wounds of soft parts and fractures of bones.

There are five different modes in which wounds may heal. 1st. *By immediate union.* Here the cut or separated surfaces, being held in close apposition, literally unite without the intervention of any new material. 2nd. *By primary adhesion.* When the surfaces are not closely joined, an adhesive, plastic material, called *lymph*, is poured out and fills up the gap between them. This lymph, either by development of the cells which it contains or by fibrillation, is converted into connective-tissue, which acts as a bond of union. 3rd. *By granulation.* Where the breach of continuity is still wider and the surface exposed to the air, the exuded lymph is more abundant, its cells multiply excessively and lie heaped in clusters, in and among which newly-formed blood-vessels ramify, till at length the well-known red, fleshy mass, called "granulation tissue," is formed. The healing process is

---

* "Warm blood in Vertebrata and aerial respiration in Invertebrata seem to be incompatible with high powers of repair of injury."—*Prof. Rolleston.*

finally effected by the development of these granulations into connective-tissue, which fills up the deeper parts, and cuticle, which
covers the surface. 4th. *By union of granulations.* When two granu-
lating surfaces are brought together and held in contact, they will
unite. This is exemplified daily in the healing of stumps after ampu-
tation. 5th. *By healing under a scab.* From the non-exposure to air
no granulation or formation of connective-tissue takes place, but the
part beneath the scab simply " skins over."

The repair of broken bones instances a very perfect kind of repair,
in that the newly-formed material imitates exactly in its minutest
details the bony tissue. The process generally differs in man
from that which is observed in lower animals. In the latter,
the interposition of reparative material, or " callus," *between*
the fractured ends is preceded by the formation of a thick
layer of the same material *around* the fragments, encircling them
completely, and called " provisional or ensheathing callus," because
it serves to keep them in place while the process of joining is being
carried on. In the former, except in the case of the ribs, no such
provision is essential, but the joining of the fragments by means of
the intermediate callus alone proceeds ab initio. The reason of this
is probably in part owing to the artificial rest which man is enabled
to obtain, and which serves him in the place of a provisional callus,
in part also to the greater power possessed by the lower animals of
forming new bone. In the ribs complete rest is unattainable, and
this callus is consequently formed. The callus, then, which is to bone
what the adhesive lymph is to the soft parts in union by primary
adhesion, gradually ossifies, passing through the intermediate stage
of either connective-tissue or cartilage according to circumstances.

Wall-Case under Gallery No. 8.	Nos. in Catalogue.
Reproduction of lost limbs	20
Repair of soft parts	15, 21
Repair of bone, *a.* in lower animals	19
„ „ *b.* in man	16, 18
Recovery from disease	17

[*See also Special Series, Nos.* 88, 92, 93, 94, 101—110.]
Consult especially, *Paget's Lectures, p.* 115 *et sqq.,* 1863

IV. INFLAMMATION.

It has been already shewn that one of the effects produced by the action of an irritant on a given tissue is hypertrophy of that tissue. Inflammation, which is also a consequence of an irritant, is in its very beginning scarcely more than hypertrophy or excessive nutritive activity, though it soon, quickly and violently in proportion to the severity of the irritant, or the constitution of the subject of it, assumes all those essentially pathological characters which make it so important a subject of study in medicine.

A right comprehension of the real nature of inflammation will be materially assisted by a proper understanding of its exciting cause. The cause is, as has already been stated, an *irritant*. This can be asserted positively for certain inflammations which are superficial and visible. By the application of a mechanical or chemical irritant to any part of the surface of the body, inflammation can be produced at will. It can not only be seen with the naked eye, but, in the transparent tissues of certain animals, it can be watched in every detail with the microscope. We are therefore justified in assuming that what can be seen without also takes place within; and that, just as by the introduction of certain poisons into the blood inflammation of certain internal organs can be produced at will, so an irritant condition of the blood, due either to the presence of abnormal substances or to an altered state of its natural constituents, may give rise to internal inflammations. Hence it may be stated, that for the occurrence of every inflammation a previous irritation is required; and that the irritant may come from without (mechanical or chemical), or from within (through the blood, and perhaps also through the nervous system).

Now supposing that a part has been thus irritated, there follow a number of changes in the tissues themselves and in the blood-vessels of the part, so complex and seemingly so simultaneous in their action, that it is difficult to say which precede and which follow. This, however, may be taken for granted, that the primary and essential nutritive disturbance is in the tissues themselves, and that the changes in the blood-vessels are secondary.

If the wing of a bat or web of a frog be placed during life beneath the microscope and then irritated (by scratching or by the application of mustard), the first change that is *seen* is contraction of the blood-vessels, with acceleration of the stream of blood that flows through them. This continues for a variable time, and is then followed by dilatation, with retardation of the current. It will be noticed that, in each little artery, the central stream of corpuscles widens, and the marginal stream of liquor sanguinis grows narrower, till the whole calibre of the vessel is crowded with red corpuscles, which acquire an adhesive property, and run together unnaturally. The stream becoming slower and slower, oscillates, and eventually stagnates. Around this centre of stagnation it will be seen that the blood-vessels are full, and their current slow; while, in an area still farther from the centre, the stream is unnaturally strong and quick, or, in other words, there is determination of blood to the part. The changes which are taking place simultaneously in the tissues themselves are essentially those of nutritive irritation. The different elements are enlarged and multiplied by the assimilation of an excessive amount of nutritive material; while the liquor sanguinis transudes through the walls of the altered blood-vessels and soaks the tissues, or, in the case of serous membranes, escapes as a free exudation * from the inflamed surface.

Any part of the body which is the seat of the above complex changes is said to be inflamed; its functions are suspended; its *nutrition is disturbed.*

In the most favourable case, when the above changes have reached their acme, there follows resolution of the inflammation ; the current

* The question of exudation has of late years been much discussed. By Virchow and his followers it is contended that there is no *exudation of lymph* properly speaking : that in internal or parenchymatous inflammations, such as that of the liver, there is no lymph whatever poured out, as is supposed, into the *interstices* of the tissues, but that the swelling and cloudy appearance of such an inflamed part are due to the taking up of an excessive quantity of nutritive material into the tissues themselves (the cells and intercellular substance): that in superficial inflammations, such as that of serous membranes, the *serum* of the blood transudes and carries with it the fibrine *which is manufactured at the very part by the irritated and inflamed tissues.* There are no doubt inflammations in certain non-vascular parts (as the cornea and cartilage), in which there is no interstitial effusion: but for the vascular tissues the absence of exudation must not yet be regarded as proven.

It may be here mentioned, that the fact of non-vascular parts being just as liable as vascular parts to the nutritive changes implied by the word *inflammation*, affords the strongest evidence that the changes in the blood-vessels are not the essential part of the process: but that the irritated tissues themselves are the primary and principal agents.

Professor Lister believes that the dilatation of the blood-vessels after an irritant is dependent on nervous action, in that it extends for some distance round the area of the centre of irritation ; that the stagnation or unnatural aggregation of the blood corpuscles is a direct effect of the irritant, and shews that the tissues and blood-vessels, having lost their functional activity, are in a manner dead: that, in consequence of this suspension of vital activity, the blood-discs run together in the blood-vessels just as they do when shed out into a cup, or examined on a slide under the microscope.

of the blood flows on; the effused products are re-absorbed, and the part is completely restored. But it often happens, and more especially in the serous membranes, that a new material is formed, either, as some believe, out of the effused lymph, which becomes organised, or, as others maintain, out of the pre-existing cells of the connective-tissue or epithelium. This material, as may be inferred from what has been already stated in the history of the healing process, is connective-tissue. It is well studied in what are called "false membranes." These are seen in their earliest stage in the form of a layer of soft substance, which at first adheres closely to the inflamed serous membrane, but may readily be stripped off; it gradually then becomes firmer, receives newly-formed blood-vessels and lymphatics, which pass into it from the organ on which it lies, till at length it becomes an organized membrane, answering to the cicatrix-tissue in the healing of wounds.

Another way in which inflammation terminates is by *suppuration*. The formation of pus is considered by some pathologists (as Paget) to be a retrograde process, by which the lymph effused under the stress of inflammation degenerates, its fluid part being converted into liquor puris, its corpuscles into pus cells. Others (as Virchow), who deny the development of cells out of blastema, and maintain that every cell springs from a parent cell, "omnis cellula e cellulâ," hold, with more probability, that pus is formed out of pre-existing tissue; that, in the deep-seated parts, the pus cells originate in the corpuscles of the connective-tissue by partition of their nuclei and of the corpuscles themselves; whereas, in the more superficial parts, as the mucous membranes and the epidermis, the pus cells are formed out of the epithelium cells by the same process of partition and multiplication of those cells; and that the fluid part of the pus is in all cases a product of the liquefaction of the intercellular substance.

There are three principal forms of suppuration. 1st. The *circumscribed*, best illustrated by a common abscess, in which the centre of an inflamed mass has suppurated, and is limited by a hard boundary of inflamed tissue. The centre may extend, and the abscess may spread, but it still continues strictly circumscribed by its hard boundary wall. 2nd. The *diffuse*. The area of the original inflammation is in this kind more extended, and, consequently, the suppuration is not limited. This condition affects more especially the cutis and loose, porous, subcutaneous connective-tissue, and is well exemplified in the so-called phlegmonous crysipelas. 3rd. The *superficial*. In this form, as has been stated, the epithelium is the probable source from whence the pus is derived. The catarrhs of the different mucous membranes are instances in point.

, The products of inflammation shew, as indeed does all tissue thus unnaturally produced, a tendency, in certain cases, to undergo the granular or fatty degeneration. This it is which constitutes the process known as " inflammatory softening" (so common in the brain and spinal cord); while it is the essential phenomenon in that important pathological condition called *ulceration*. The term " interstitial absorption" is used to signify the degeneration and removal of deep-seated parts, as distinguished from ulceration, which is employed to express the same death in exposed or superficial parts. In both there is granular or fatty degeneration of inflamed tissues, but in the former the changes are concealed from view. For instance, in the case of an abscess which makes its way to the surface, the intervening tissues gradually disappear, degenerate, and are removed either by absorption into the blood or by breaking up into a granular detritus which mingles with the pus : while in the case of superficial ulceration the process is open to inspection, and the microscope shews that the granules or molecules are cast off, at least in greater part, from the discharging surface.

Ulceration is therefore, in a few words, molecular death of tissue either on the surface of the body or on the mucous tract, and ejection of the dead particles in what is called the discharge of the ulcer, which may consist in part of ill-formed pus, but is composed principally of this granular débris.

Lastly, Inflammation may end in *mortification*. This is to a part of the body what death is to the whole. The term is properly employed to signify the death of a portion of the body large enough to be seen with the naked eye. Mortification differs, therefore, from ulceration only in degree, and under certain circumstances approaches so nearly to that process that it can hardly be said where the latter ends and the former begins. For instance, in the morbid condition known as " phagedæna," the two processes, molecular death and death en masse, go on together.

The causes of mortification are, 1st. Violent mechanical or chemical injuries, such as destroy the tissues at once. 2nd. Causes which act less directly by interfering with the circulation of a part and killing it by depriving it of its nutrition. Among these stands inflammation the first. It has already been shewn that two of the changes which take place in inflammation are stagnation of the blood and subsequent degeneration of the tissues. If, now, the inflammation be very severe, or, more particularly, if it attack a system already enfeebled, the stagnation becomes excessive, and, instead of degeneration, death results. In such a case, the dying tissues contain their blood and products of inflammation, and the part putrefies: the mortification is then said to be " moist." Again, the flow of blood to a part may be checked by some obstruction in the artery which supplies it, either through disease in the coats of the artery causing a gradual narrowing of its calibre and coagulation of the blood in it, or through "embolism" [i. e. the transmission of a detached fragment of fibrine, or other substance, from one part of the circulation to another], by means of which a fragment is impacted in an artery or capillary which is too small to let it pass. In this case the part supplied by the occluded artery may undergo the moist form of gangrene, but may also gradually wither and die, just as the branch of a tree dries up, when the sap no longer circulates in it. This "dry" gangrene is especially apt to occur in parts where there is comparatively little soft tissue, as in the feet and hands. Stagnation of the blood and mortification may also take place in consequence of constriction or strangulation of a part, as in hernia or intro-susception. (See *Prep.* 422.) 3rd. Certain abnormal conditions of the blood, either of spontaneous origin or induced, may bring about mortification. As instances, may be cited, *a.* Carbuncles and boils, which are localised inflammations attended with mortification: *b.* Sloughing of certain tissues (as the cornea) in starvation: *c.* Mortification of different organs in consequence of improper food (as ergot of rye).

The above effects of inflammation in the soft parts are repeated in bone under the changes denominated " caries" and " necrosis" (see *Preparations* 116—121; and 127—136): the former being the counterpart of ulceration, the latter of mortification.

Wall-Case under Gallery No. 8.

INFLAMMATION

	Nos. in Catalogue.
Of mucous membrane	22
Of lung tissue	23
Of serous membrane with " effused lymph "	24, 26
Effused lymph organised, or " false membrane"	25, 27, 28
Formation of new vessels in false membrane,	
arteries	29 — 31
veins	32
lymphatics	33 — 36

See *Paget's Lectures,* 1863, *pp.* 218–352; *Simon, Lect. on Gen. Path.* 1850. *See also Holmes's Surgery, Vol. I. p.* 1; *Virchow, Handbuch der Spec. Path. und Therap. Vol. I. p.* 46; *Virchow's Cellular Pathol.,* translated by *Chance, p.* 384; *Lebert, Path. Anat. Vol. I. pp.* 27–61; *Aitken, Science and Practice of Medicine,* 4th *Ed. Vol. I. p.* 64; *Lister, Philos. Trans.* 1858, *p.* 645; *Wharton Jones, Guy's Hosp. Rep.* 1851, p. 1; *C. O. Weber, Die Entwickelung des Eiters—Virchow's Archiv.* 1859. *For the earlier history of the subject, see W. P. Alison, in Practice of Medicine, Vol. IX.; Library of Medicine, Vol. I.; and Thomson on Inflammation.*

V. TUBERCLE.

A tubercle is a particle, 'knot,' or granule, either isolated or occurring in clusters, having, according to its age, a grayish or yellowish colour, met with in nearly all organs of the body, but typical in serous membranes. Such a particle selected for examination from a serous membrane, pleura or peritonæum, (though it may be equally well taken from any organ), is, when young, gray, semi-transparent, of about the size of a millet-seed, (whence the name of "miliary" tubercle), slightly raised above the level of the serous layer, but evidently lying just beneath that layer, which is continued smoothly over it. At this stage of its life it is seen with the microscope to be made up of small cells very like those of lymphatic glands, round, varying somewhat in size, but for the most part rather smaller than the white blood-cell. A small shining nucleus is contained in the cell, sometimes, but not always, inclosing a nucleolus, and surrounded by a faintly granular, cellular substance. Besides the above, some larger cells, containing two or more nuclei, and abundant free nuclei are also observed. Between the cells or nuclei is a fine network of connective-tissue, with, occasionally, a few blood-vessels, not newly-formed, but belonging to the tissue in which the tubercle is developed: for in tubercle there is no new development of blood-vessels proper to the growth, as there is in most tumours.

Of all pathological products tubercle is that which soonest degenerates; its most striking character being its early tendency to decay. The gray stage is, therefore, comparatively short-lived, and is soon succeeded by the yellow. Yellow tubercle has been long, and is still, considered by many to be a distinct form, yellow from the very beginning of its life, and possessed of a greater tendency to infiltrate the tissues and spread to adjacent parts than the gray. The truth, however, is this—that there is only one form of tubercle, which, in its early life, is gray, and which, from its ready tendency to undergo the fatty or granular degeneration, soon becomes yellow; the change of colour being simply due to this fatty degeneration, which, commencing in the centre, gradually extends to the whole of the particle. The microscope now shews cells irregular or angular in form, shrunken or withered, with no distinct cell wall, containing fatty granules, and, rarely, one or two nuclei; abundant free molecules or granules, and irregularly-shaped nuclei. At a still later period the whole knot softens and breaks down into granular débris. Such is the life and such the death of each individual tubercle, and were the little knot to remain thus isolated, its nature would not be so deadly as it is. But others, similar to itself, are apt to form in indefinite number around it, and produce by an agglomeration of individual tubercles the large tubercular masses which may involve in destruction whole organs of the body and sweep away whole tracts of mucous membrane. Such a tubercular mass, then, whether it occur in the form of a tumour (as in the brain, see *Prepar.* 771 to 774, 785, 6) or in that of a so-called deposit (as in the lung), is really nothing else than an aggregation of little knots or tubercles proper. The degeneration which, when the tubercle is isolated, terminates in death of the individual tubercle, when the tubercles are clustered, results in death en masse: and the consequent destruction of the organ involved must be in direct proportion cæteris paribus to the number of tubercles in the cluster. It will also be understood that, if the part diseased be a free surface (as a mucous membrane), the disintegration and softening of the tubercular mass will give rise to the formation of an ulcer, reminding us of the ordinary process of ulceration, but differing from that process in that the base and sides of the ulcer are walled in by the remaining tubercular matter instead of by the natural tissues. In like manner it is clear that if the seat of the tubercle be the parenchyma of some organ, as the lung, a cavity or " vomica" must result, in which are contained the granular remains of the softened tubercle and of the broken-down tissue involved by the tubercle in destruction.

The seat of tubercle is, according to the order of frequency, first in the lungs, then in the intestines, lymphatic glands, larynx, serous membranes, pia mater, brain, spleen, kidneys, liver, bones and perios-teum, uterus, testicle (especially epididymis), prostate gland, vesiculæ seminales. Among glands, the pancreas, parotid, mammary, and thyroid; among tissues, the muscular—are the most rarely affected with tubercle.

The cause of tubercle is, more frequently inheritance than is the case with any other affection. It is still a matter of doubt whether it be in the blood or in the tissues themselves that the inherited tendency is localised, but this fundamental truth cannot be challenged—that a disposition to this peculiar disease is conveyed to the embryo by the parent; and that this *pre*disposition, in spite of the innumerable changes that take place during development and growth, remains as a part of the individual; though dormant, it is ready, as occasion offers, to manifest itself in him, often in the same organ, often at the same period of life, as it shewed itself in his parent. But at the same time that due weight is attached to this most potent predisposing cause, it must be borne in mind that the direct irritant which excites the disease is often local and external, and that, therefore, as regards its cause, tubercle must rank with those tumours which depend generally on constitutional influences, but, are often a consequence of local irritation. The frequency with which tubercular disease, where it affects the mucous tract, follows catarrh of that tract, and, when it attacks the lung, follows catarrh of the bronchi, bears out the truth of what is here stated. A knowledge of this circumstance is also of consequence to the physician, as enabling him to warn those who are hereditarily predisposed to tubercular disease against exposure to cold, to excessive fatigue, irregular habits, and all analogous conditions which can be justly counted among exciting causes of the disease.

The origin of tubercle is in the connective-tissue, or in one of the same family—in marrow, fat, bone, and, more rarely, epithelium. It begins by enlargement, partition and multiplication first of nuclei then of cells, which group themselves into the particles or masses previously described.

The termination of tubercle is most frequently by granular degene-ration, and the consequent formation of ulcers on free surfaces, of cavities in the interior of organs. But tubercle may, under favourable circumstances, heal. First, there is a high degree of probability that it can be completely absorbed and so removed. For it is not uncommon to find after death cicatrices, in which no traces of tubercle remain, in such parts of the body as the apices of the lungs, where tubercle has its favourite seat. When this occurs in persons who have been known during life to suffer from all the signs, both general and physical, which are characteristic of the tubercular disease, the proof may be counted to be complete. A favourable termination is, however, more often brought about by drying up of the tubercle and its conversion into a chalky, inactive mass, which remains in situ, but has lost its malignant properties.

It remains to be stated that even on careful microscopical examina-tion of tubercle at an early period, it cannot be certainly dis-tinguished from cancer at the same stage of its growth. But whereas cancer goes a stage further in development, and has great, often unlimited, powers of growth; tubercle has no capability of improving or increasing itself, but is henceforward retrograde instead of progressive. Tubercle, like cancer, is prone to affect lymphatic glands secondarily (i.e. the mesenteric glands in primary tubercular disease of the intestines, or the bronchial glands when the primary disease is in the lungs); it is apt to disseminate itself, as if by metastasis, (as to the liver, kidneys, etc. from the lung, and to the lungs and pleura from the kidney); and it sometimes assumes the exact form of a tumour (as in the brain, where it can often be with difficulty distinguished from cancerous and other tumours).

The affinities of the elements of tubercle with other tumours demand especial attention.

Young Tubercle, called also gray or miliary Tubercle—

Old Tubercle, called also yellow or crude Tubercle—

1. In an early stage of degeneration.

        of the lung 634, 649
        „   „   glands ... . 533 to 535
        „   „   supra-renal capsules ...... 547 to 547, *b.*
        „   „   kidney 718
        „   „   brain .... 771 to 774, 785, 786

2. In an advanced stage of degeneration, softened and broken
   down, so as to form

    A. Ulcers

        of the larynx 670
        „   „   trachea 675
        „   „   intestine .... 379 to 388, 395, 396
        „   „   ureter .... 715 to 717

    B. Cavities

        of the lung ..... 58 to 60, 648, 650 to 659, 662
        „   „   kidney .... 696, 697, 712 to 717

Tubercle cretified and inactive—

        of the lung .... 640 to 647
        „   „   glands 522, 523, 678

Virchow, Cellular Pathol., Lect. 20; *Die Krankh. Geschwülste,*
Vol. II. Lect. 24; *Paget, Op. Cit., p.* 812; *Lebert, Op. Cit.,*
p. 328; *Holmes's Surgery, Vol. I. p.* 336; *Rokitansky, Op. Cit.,*
p. 292.

VI. TUMOURS.

The closeness of the link which connects hypertrophies with
tumours has been already alluded to. Certain organs of the body,
as the prostate and thyroid glands, are not only subject to general
hypertrophy, as before indicated, but are often the seat of localised
hypertrophies in the form of limited pendulous outgrowths of their
tissue, which in appearance, in form, and in danger to adjacent parts,
have such near affinities with tumours, that they must be regarded as
occupying a neutral ground between hypertrophies on the one side,
and the class of morbid products now to be discussed, on the other.

So wide is the difference between the simplest form that a tumour
can assume and the compound masses which, at the other end of the

scale, may infest the body—masses which, in their complexity, can imitate whole organs or systems of the body,—that it is undesirable to attempt any definition of the word " tumour." But all have, in a greater or less degree, peculiarities which mark them, characters unlike those to be observed in the natural development and growth of the body. A tumour is strictly a part of the body, and is not only in connexion with it, but proceeds from it. It derives sustenance from it and is subject to its laws: but it grows " with no seeming purpose" or definite end, deviates in growing ever farther and farther from the normal shape and typical plan of the parts around, nay, in many instances, bears not the slightest resemblance in structure to those parts, appears to be useless, is generally destructive, often deadly.

The same cause which we have seen to be essential to hypertrophy and to inflammation, is essential also to all tumour formation—viz. the action of an *irritant*. The analogy is still further strengthened by the way in which the irritant acts. For first the irritant may come from without, and the cause is said to be *local*. Secondly, the irritant may come from within, and the cause is said to be *constitutional*. With respect to the first class, or *local* tumours, it is certain that they are often the result of a blow or some kindred mechanical injury. With respect to the second class, or *constitutional* tumours, (the largest group,) it is supposed that either the blood or the solid tissues have a special predisposition, inherited or acquired, to such formations.*

When once the irritant has acted, the first and most essential change begins, as in inflammation, in the elements of the irritated tissue. The tissue which most frequently undergoes this change is the connective tissue, and next to it the tissues most nearly allied to it—the medulla of bone, and, more rarely, bone and cartilage. These elements begin at once to take up an increased quantity of nutritive material, they swell and the cells begin to multiply. First the nuclei, and next the cells, undergo division or partition, till, by frequent repetition of the process,

---

* Even here, however, the exciting cause can be traced, in some instances, to external influences. Thus, Paget states that about a fifth of those who have cancer (the typical constitutional tumour) ascribe it to an injury. Virchow, too, points out how the female breast and the stomach, two organs which are more especially the seat of active changes, and which are also exposed to the action of external agencies, are especially liable to be the seat of cancer. The tendency of advancing knowledge seems to be to narrow the circle of constitutional, and widen that of local, causes.

masses or clusters of cells are seen to be aggregated together, and the embryo of the future tumour is formed. Up to this stage of development it is impossible to say what kind of tumour is in process of formation: cancer, tubercle, connective-tissue tumour, all have the same appearance. There is, no doubt, some special property already present in the groups of cells which determines their future course of development, just as in the cellular stage of the embryo the different cells must be endowed with a force which compels them to develope into some particular tissue: but such properties and such a force are invisible ; the cells appear to be the same. From this time forward the proper characters of the tumour, whatever it is to be, begin to shew themselves. In some there is scarcely any advance beyond this simple cellular or " granulation *" stage. In others there is a little further progress, and the growth assumes the form and structure of some simple tissue of the body. In others, again, development is on a still higher scale, and some complete organ (e. g. gland), in its compound form and variety of elements, is imitated. Or, lastly, the tumour may contain an exact representation of some still more complex normal structure, as perfect skin, with its glands, hairs, etc. In the formation of some simple tumours the above plan of development is not followed out, but the tumour elements assume *at once* the same form as those of the mother element from which they spring, without passing through this "granulation stage." The most important, however, because the most destructive tumours are formed as above.

Concerning the characteristics of tumours it may be stated that in no tumour can elements be found which are altogether dissimilar from some that normally exist in the body either in its embryo or in its adult state. It has been held that the microscope is able to pronounce on the nature of a given tumour, by the presence of cells special or peculiar to the morbid growth and unlike any normal structure. Hence it is that tumours are said to be either "homologous" or "heterologous," according as they, in their composition, resemble or differ from tissues which are naturally present in the body. But if it be the case that every pathological cell has its physiological prototype either in the embryo or in the adult body, then no tumour can be in the above sense heterologous. It is better to designate those tumours as heterologous whose structure differs

† Name given by Virchow, because of the very close resemblance that the growth bears to the ordinary granulations which form in the healing of wounds, ulcers, etc.

from the mother-tissue in which they grow (e. g. a cartilaginous tumour in the testicle, a bony tumour in the lung), and to apply the term homologous to tumours whose structure is identical with that of their mother-tissue (e. g. a fatty tumour in fat, a bony tumour in bone).

The majority of tumours have blood-vessels and lymphatics of their own, but in all cases directly continuous with the vessels of the body, so that the tumour may be injected from the main artery in its neighbourhood.

A large proportion of tumours are only harmful to the part of the body in which they grow, and when removed by operation never more appear. They may increase in size; but they remain fixed to the spot where they were first developed, and are only dangerous by the pressure which they exercise on surrounding parts. Their hurtfulness bears therefore a direct proportion to the importance of the organ in or near which they are seated. All such tumours are said to be " innocent or benign." But there are several kinds which reappear in the place from whence they have been removed by operation, and which not only increase locally, but disseminate themselves, producing secondarily, either by diffusion of their juices or by positive transportation of their elements, tumours similar to themselves in different parts of the body. They are dépôts from whence disease emanates in every direction. The agents of this diffusion or transport are first the lymphatics, in which case the nearest group of lymphatic glands become secondarily affected; secondly the blood-vessels, when all the organs of the body are in danger of being involved. Such tumours are said to be " malignant." It may be laid down as a rule that the more heterologous a tumour is, the more likely it is to be malignant. Again, the more soft and succulent, the more dangerous. On the other hand, homologous tumours are for the most part benign ; and those which are dry and juiceless are generally, though not always, benign *.

* Allusion has been made, under the head of Inflammation, to the blastema theory, as contradistinguished from the cellular. The same two theories are here again seen to clash. The one, maintains that the first beginning of every tumour is a free blastema or plastic fluid which exudes from the blood, and out of which the tumour elements are developed, and refers necessarily the origin of all tumours to the blood, and illustrates well what is called the " humoro-pathological doctrine." The other, regards the tissues themselves as the absolute starting-point of all tumours, and denies the existence of any free exudation, and only allows of an excessive transit of nutritive material from the blood directly into the very substance of the irritated tissues. It is certain that what can be seen with the microscope, in the examination of embryo-tumours, strongly supports the latter theory.

Tumours should be classed according to their anatomical characters. They are either *cystic* or *solid*.

1. CYSTIC TUMOURS.

A *cyst* is a sac or bag in which are contents of various consistence. The sac itself is composed mainly of connective-tissue, and is often lined with epithelium. It is formed either as a consequence of an unnatural dilatation of some sacculus or duct naturally present in the body (e. g. a cyst, of the ovary, from dilatation of a Graafian vesicle; of the kidney, from dilatation of an obstructed uriniferous tubule; of the breast, from an obstructed lactiferous duct, etc.): or out of connective-tissue by an enlargement and fusion of the spaces in that tissue (e. g. bursa in the subcutaneous connective-tissue) : or by monstrous growth of pre-existing cells or nuclei (e. g. certain cysts in the kidney and cysts in bones).

The contents are either fluid or semifluid or solid. Of the purely fluid, blood and serum offer the best instances. Of the semifluid, the thick, ropy, viscid substance, white or tinted of different colours, met with so commonly in ovarian and thyroid cysts, and the thick synovia-like contents of bursæ, form the best illustrations. Of the solid, the masses of epithelium and cholestearine crystals in the so-called sebaceous or cutaneous cysts, the complete skin, bones, or teeth in some ovarian cysts, and the solid masses which grow into and gradually fill up the cavities of certain cysts (e. g. of the breast, in the mass known by the name of "sero-cystic sarcoma") are good examples.

Cysts are either simple or compound. The simple, which are also called barren, are single, i. e. consist of but one solitary sac and its contents. The compound are proliferous, i. e. they have the power of producing from some part of their walls secondary or daughter cysts. This power is sometimes very highly developed, so that the secondary cysts may be seen, in very great numbers, growing into and partially filling up the cavity of the parent.

Simple or barren Cysts
{
Of the Ovary 51, 908, 911, 912
Of the Kidney ... 698–703, 706, 707, 719
Of the Fallopian Tube 911, 916–921
}

Compound or Proliferous Cysts.   Of the Ovary 910

2. SOLID TUMOURS[*].

1st. *Fatty Tumours.*

Most common seat: Subcutaneous fat, particularly of the trunk.

Structure, identical with the ordinary adipose tissue; the connective-tissue which pervades them varying much in quantity.

Characters: Their consistence bears a direct relation to the quantity of connective-tissue they contain: painless as a rule: single, as a rule, but occasionally multiple, in which case all are in the same, subcutaneous, tissue: very common: innocent: homologous.

Of the subcutaneous fat 52

2nd. *Fibro-cellular Tumours.*

Most common seat: Superficial, in scrotum, labium or wall of vagina; deep-seated, between the muscles.

Structure: To the naked eye, a tissue which resembles the ordinary connective-tissue either in its embryonic or in its more perfect state, generally soaked with much serous or synovia-like fluid; with the microscope, nucleated cells mostly caudate, fusiform, or stellate; free nuclei; bundles of fine undulating fibres.

Characters: Firm but elastic to the touch, on section shew a " shining, succulent, yellowish, basis-substance, intersected by white lines:" grow rapidly for the most part, and reach a great size: rare: generally innocent, but, if very succulent and containing many nuclei and cells, tend to return in situ after removal (recurrent): homologous.

3rd. *Fibrous Tumours.*

Most common seat: Uterus, bones (especially the jaws), nerves, subcutaneous tissue, connective-tissue near joints, sheaths of tendons.

---

* This classification will be found to agree not with the Hunterian Catalogue, but with Mr. Paget's Lectures 21 to 35. These Lectures should be considered by the student as the key to this portion of Pathological Science.

Structure: To the naked eye, fibrous tissue with a manifold plan of arrangement; with the microscope, the same as fibrous tissue, with an admixture of elongated cells, and, in some cases (as in uterus), perfect organic muscular fibres.

Characters: Round or oval in shape, and generally invested by a capsule: firm and tough, more rarely soft and succulent: sometimes, when old, calcified: growth most often slow, but may reach a great size: very common: as a rule innocent, but in certain cases, when soft and containing an unusual quantity of cells, malignant: homologous.

4th. Cartilaginous Tumours.

Most common seat: Bones; more rarely soft parts, as testicle, parotid gland.

Structure: Ordinary fœtal cartilage with more or less fibro-cartilage, fibrous tissue, bone; with the microscope, the ordinary appearances of the above tissues, the cells being often irregular in shape.

Characters: Irregular in shape, generally lobed: as a rule hard, but occasionally softer and almost diffluent: differ much as to size and rate of growth: generally innocent, but sometimes, especially when soft, malignant: generally homologous, but occasionally (when occurring in soft parts) heterologous.

5th. Myeloid Tumours.

Most common seat: In or on bones.

Structure: The distinguishing feature of these tumours is the large,

many-nucleated cell of which they are in part composed, and which is a constituent of the fœtal marrow.
These myeloid cells are mixed with caudate cells, exactly resembling those in the fibro-cellular tumours,
and with free nuclei.

Characters: Growth usually slow and painless: as a rule innocent:
when enclosed *in* bone, well-defined, spherical or oval
in shape; when growing *on* bone, irregular and lobed:
consistence firm and " fleshy," but brittle; more rarely
soft: on section often marked with blotches of different colours.

6th. *Osseous Tumours.*

Most common seat: Bones, from which they are outgrowths; very
rarely soft parts.

Structure: An exact imitation of cancellous or compact bony tissue.

Characters: Merely ossified cartilaginous tumours in many cases;
many of them (especially the hardest and most ivory-
like) grow directly from the fibrous tissue: some cancellous and softer; others very compact, like ivory:
generally single, but occasionally multiple, in which
case all the tumours are in the same tissue: innocent:
homologous; very rarely heterologous.

Exostosis 138, 139, 221

7th. *Glandular Tumours.*

Most common seat: In or near glands, whose structure the tumour
imitates (mammary gland typical, labial, prostate,
thyroid).

Structure: An exact imitation of the gland, generally in its rudi-
mentary state, near which the tumour grows; with
microscope, abundant gland cells grouped or clustered
in acinous form; connective-tissue more or less abundant; in some, muscular fibres.

Characters: Round or oval in shape: invested with a capsule: on
section, lobed, with a decidedly glandular aspect; some
white and firm, others yellowish, softer and more
succulent: as a rule, painless and of slow growth:

innocent: generally homologous, but, when growing independently of glands, heterologous *.

Of the mammary gland 56a

8th. *Vascular or Erectile Tumours.*

Most common seat: Cutaneous and subcutaneous tissue, principally of scalp, face, and trunk, but may be met with in internal organs, as the liver.

Structure: Agglomerations of dilated arteries, veins, or capillaries.

Characters: Round or oval, more or less flattened: generally red, but the colour does not always shew through the skin: rate of growth uncertain: as a rule date from infancy (nævus the type).

In studying the above tumours, which are by some writers grouped together under the head of "innocent tumours," it will be remarked that a few of them are, under certain circumstances, decidedly malignant. The *Cancerous Tumours*, which come next in order, are, on the other hand, commonly spoken of as "malignant tumours" par excellence, in that their natural character is malignant. But yet no absolute limit can be drawn between the innocent and malignant tumours; the character of malignancy being, though the main, by no means the exclusive, property of the latter. Again, the cancerous tumour is always heterologous, but so are many of the non-cancerous. Lastly, though the anatomical elements of cancers are unlike the elements in the midst of which they lie, yet they have their prototype in the body just as much as is the case with the most innocent tumours.

The peculiarities which especially belong to the cancers may be thus enumerated. 1st. That their elements are infiltrated in the interstices of the natural elements, and bear no resemblance to those which surround them. 2nd. That they are of a very low organisation, their materials being never "elaborated into an enduring structure." 3rd. That they are, in the widest sense of the term, malignant. The question may be asked, Can a cancer be certainly distinguished by means of the microscope? To this it may be

---

* These glandular tumours differ from the hypertrophied outgrowths of glands already alluded to, in that they are distinctly separated from the gland by a capsule, or are quite disconnected from the gland, and merely lie in its neighbourhood.

answered, that it is not possible to determine by the microscopical characters of one cell in any given tumour whether that tumour is cancerous or not; but by taking several points into consideration— the way in which the tumour has grown, whether it is infiltrated into and not clearly separable from the parts where it grows, the abundant presence of cells with large nuclei, the disorderly grouping of those cells,—the nature and probable consequences of a given tumour may be with accuracy determined. *The mere form of the cells which compose a tumour affords no positive indication of the nature of that tumour.*

***

9th. Hard or Scirrhous Cancerous Tumours.

Most common seat: Typical in the female breast; in the stomach and large intestine, and therein especially the rectum, sigmoid flexure, cæcum.

Structure: A growth of cells beginning in the connective-tissue of the affected organ, and gradually spreading so that the cells become infiltrated into the interstices of the organ: hence the tumour which results consists in part of the cancer-cells, in part of the original tissue of the diseased organ, which has not been destroyed by the new growth, and which remains to form what is called the " stroma" of the cancer. The quantity of this stroma, which consists mainly of fibrous tissue, varies much according to the ravages which the cell-growth, the essence of the tumour, has made. The cells are nucleated, of different shapes, the prevailing shape round or oval, from $\frac{1}{1000}$ to $\frac{1}{700}$ inch in diameter. The nucleus unusually large, oval, clear, and defined, sometimes single, sometimes in groups of two or three, containing a large and shining nucleolus, fills up a considerable part of the cell, and is surrounded by granular cell-contents; the whole sometimes enclosed in a distinct cell-wall; but in other cells there is no such boundary, the nucleus lying, apparently, imbedded in a portion of granular substance which has no regular outline, and seems to have been detached as a fragment: free nuclei, often in considerable numbers, are also seen. In tumours that have been kept long or are degenerating, a certain quantity of fatty and granular matter is always present.

Characters: The hardness of this tumour is its most noticeable feature: smallness of size and density of structure also characterise it: it is prone to ulcerate, the base and margins of the ulcer being cancerous, and shewing, consequently, no disposition to heal: lowly vascular: slow of growth as a rule: highly malignant.

10th. *Soft or Medullary Cancerous Tumours.*

Most common seat: The bones and muscles, especially of the extremities; the testicle, the eye, the stomach and intestines. More rarely in the liver, lungs, kidneys, uterus, and breast (where it is as rare as scirrhus is common). It is common in lymphatic glands, but in them is for the most part secondary.

Structure : Sometimes encapsuled, more often infiltrated : a fine stroma of connective-tissue and blood-vessels, holding in its meshes myriads of cells and soaked with fluid; the toughness of the tumour depends directly on the greater or less quantity of the stroma: the description of the cells in scirrhous tumours may be taken for those in the medullary; the chief difference being that, owing to the much more rapid growth of the latter, there is a greater variety of shape (many caudate, fusiform, and oat-shaped) and a greater abundance of free nuclei.

Characters: Softness and juiciness are predominant: they are very rapid in growth, and reach a great size: very vascular, and prone to bleed (hence the term " fungus hæmatodes") : from their rapid growth tend to push before

them the surrounding parts, as well as to creep in and
between the parts, like scirrhus: in appearance brain-
like: highly malignant, the very type of malignant
tumours.

11*th. Melanotic Cancerous Tumours.*

Most common seat: Skin (especially moles), eye.

Structure: They are medullary cancers which contain pigment granulations either free or enclosed in the cancer-cells.

Characters: Their blackness at once distinguishes them from all
other forms: very malignant, shewing an unusual tendency to multiply.

12*th. Epithelial Cancerous Tumours.*

Most common seat: In or beneath the skin and mucous membranes
(especially of the lip, tongue, œsophagus, penis, scrotum, uterus, labia, nymphæ).

Structure: Cells of tesselated epithelium, either grouped irregularly
or wrapped together in regular layers (like those of an
onion) so as to form small globular bodies visible even
to the naked eye, the "laminated capsules" or "globes

THE following References to the Hand-Catalogues and to the corresponding Preparations which illustrate the Series of Special Pathology may be convenient to persons studying the Collection of General Pathology :—

Wall-Cases 27, 28, 29, 30.
MS. Catalogue, Vol. 12.
Nervous system (nerves and sense organs) ....818—898

Wall-Cases 30, 31, 32, 33.
MS. Catalogue, Vol. 13.
Organs of generation and breast ....899—985

Wall-Cases 33, 34, 35, 36, 37.
MS. Catalogue, Vol. 14.
Malformations, monsters, parasites ....986—1051

Gallery Flat Cases 38, 39, 40.
MS. Catalogue, Vol. 15.
Calculi.

Cabinet No. 7.
MS. Catalogue, Vol. 16.
Microscopic preparations.

Cabinets Nos. 8, 9.
MS. Catalogue, Vol. 17.
Instruments of diagnosis.